Measuring the Weather

Precipitation

Alan Rodgers and Angella Streluk

Heinemann Library
Chicago, Illinois

Published by Heinemann Library,
an imprint of Reed Educational & Professional Publishing,
Chicago, Illinois
Customer Service 888-454-2279
Visit our website at www.heinemannlibrary.com

Design by Storeybooks
Originated by Ambassador Litho Limited
Printed in Hong Kong/China

07 06 05 04 03
10 9 8 7 6 5 4 3 2 1

Library of Congress Cataloging-in-Publication Data

Rodgers, Alan, 1958-
Precipitation / Alan Rodgers and Angella Streluk.
p. cm. -- (Measuring the weather)
Summary: Provides an introduction to the different types of precipitation and explains how precipitation is related to weather.
Includes bibliographical references and index.
ISBN 1-58810-688-8 (HC) -- ISBN 1-40340-128-4 (PB)
1. Precipitation (Meteorology)--Juvenile literature. [1. Precipitation (Meteorology)] I. Streluk, Angella, 1961- II. Title.
QC920 .R64 2002
551.57'7--dc21
2002004018

Acknowledgements
The Publishers would like to thank the following for permission to reproduce photographs:
Bruce Coleman Collection, p. 4; Eye Ubiquitous, p. 7; AFP, p. 9; Trip/H. Rogers, p. 10; Corbis/Dean Conger, p. 12; Heather Angel, p. 13; Stefan Streluk, p. 14; The National Met Office/W. A. Bentley, p. 16; Alan Rodgers, p. 17; FLPA, p. 18; Science Photo Library, pp. 19, 20, 21, 26; The National Met Office/M. Grinnell, p. 24; Robert Harding, p. 25; Popperfoto, pp. 27, 28; Pictor, p. 29.

Cover photographs reproduced with permission of Telegraph Colour Library and Tudor Photography.

Our thanks to Jacquie Syvret of the Met Office for her assistance during the preparation of this book.

Some words are shown in bold, **like this.** You can find out what they mean by looking in the glossary.

Contents

Water Everywhere!

Did you know that the oceans hold about 94 percent of the world's water? These oceans play an important part in our weather. The weather is like a giant machine. Together, the Sun and the oceans drive this weather machine on Earth. The weather causes the water on Earth to change from liquid into **water vapor.** In cold places, it freezes into a solid. These changes form part of the **water cycle.** The movement of water as rain, snow, and hail is all part of the weather process.

Water is very important for all living things. Not enough rainfall is bad, but too much is just as much of a disaster. **Meteorologists** therefore spend a lot of time studying water in its different forms.

Looking at the Earth from space shows the large proportion of the planet that is covered by water.

Clouds move water from one place to another. They are made of millions of gallons of **condensed** water and swirling air. Not all clouds produce rain. Even when they do, not all of it reaches the ground. Looking at the different types of clouds can be very interesting and informative.

Weather and climate

Weather and **climate** are not the same. The term "weather" is used to describe what happens from day to day (for example, it might rain or be sunny). It also describes those events that are unusual and unexpected (such as a violent storm). **Data** is collected for many years to work out the pattern of the weather in a particular area. This weather pattern over many seasons makes up that area's climate.

Meteorologists use symbols to represent different types of weather. These symbols are recognized all over the world, so that the meteorologists can share data. The symbols used by professional weather forecasters are often simplified for use on television and in newspapers.

Be careful!

Do not look directly at the Sun when studying the weather. Also, never take shelter under trees during a thunderstorm, because they could be hit by lightening.

These are the internationally understood symbols used by meteorologists to show different types of **precipitation.**

The Water Cycle

The **water cycle** is the name given to the system in which water moves and changes between its different forms. When we see it moving as it rains, we are only seeing a small part of the water cycle. Water often changes its form. Sometimes it will be **water vapor** in the **atmosphere.** Sometimes it **condenses** and becomes liquid again. Most of the world's water is in the oceans. Almost all of the rest is found in the **polar regions,** where it stays mostly as ice.

The water cycle works like this. The sun **evaporates** the moisture from the ocean, lakes, the land, and plants. Eighty-five percent of the moisture in the atmosphere comes from the oceans. The rest evaporates from plants and moist lands such as swamps. The evaporated water goes into the atmosphere. At cooler temperatures, the water vapor condenses into tiny **droplets** of water, creating clouds.

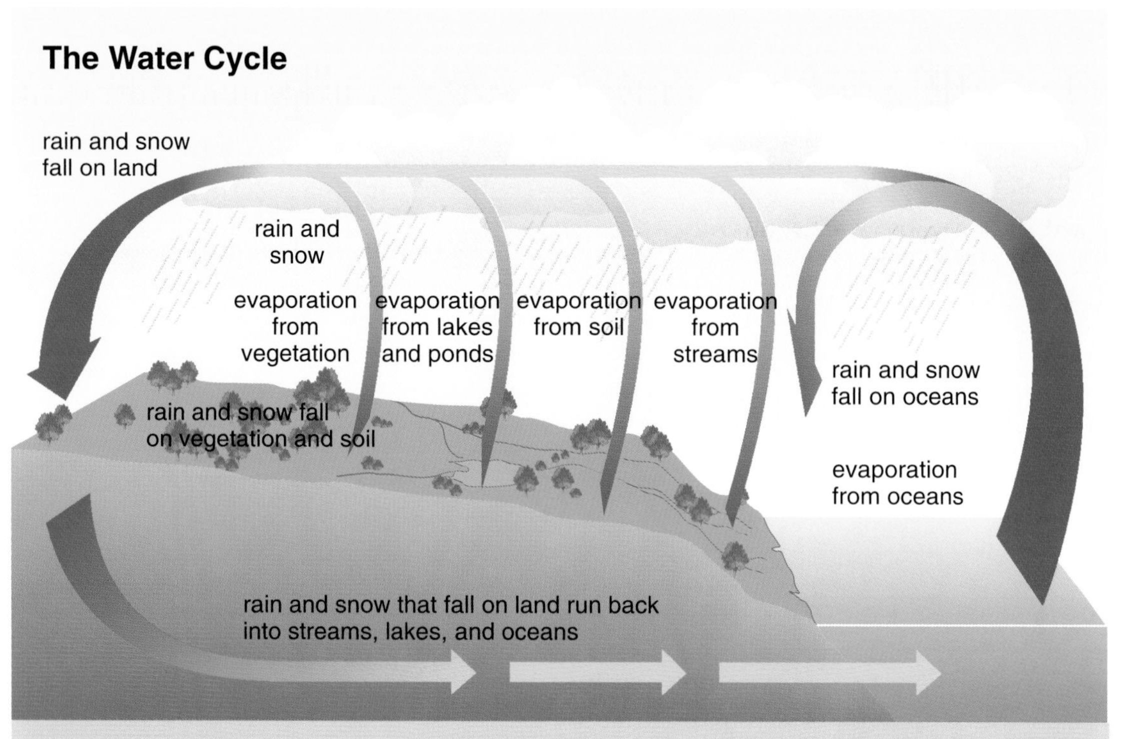

This diagram shows the water cycle. When rain or snow falls, some of the water evaporates almost immediately from wherever it lands. Other water runs into rivers, lakes, and oceans before eventually evaporating. The evaporated water enters the atmosphere, where it condenses to form clouds. Soon it will fall as rain or snow once more.

Clouds release their water in certain conditions. When clouds are blown over higher land, or warm air forces them to rise, the water can fall as rain. If it is very cold, it will fall as hail or snow. Some rain falls in the sea, and a large proportion of rain falls on land sloping up from the coast.

Rain that falls onto land eventually makes its way to streams and then to larger rivers. It eventually returns to the sea and the whole water cycle begins again.

Hydroelectric power

Heavy rain falling on hills and mountains can produce fast-flowing rivers. Many of these rivers have dams built across them with **hydroelectric** power stations inside them. These power stations use the water to produce electricity. The electricity produced in this way is environmentally friendly because it does not use up world resources, such as oil.

Although dams provide environmentally friendly energy, they are not always good for people who live close to them. When a river is dammed, all of the water that used to flow away now forms a large lake. If people are living in the area that will soon be covered by the lake, they will have to move.

Why Is It Raining Here?

The proper name for water that falls from clouds is **precipitation.** Water can fall in various forms, including **drizzle,** rain, **freezing rain,** sleet, and snow.

In order to produce rain, tiny molecules of **water vapor** need to form around very small particles of **matter.** These particles are usually tiny bits of dust, pollutants, or sea salt. If the **atmosphere** were completely pure, with no pollution or dirt, it would never rain! Combined with the water, the particles form **droplets.** These droplets do not fall right away, as they are so tiny that air currents can keep them suspended in the air. The air within a cloud is always moving, and carries water droplets and ice crystals within it. During all this movement, many ice crystals and water droplets collide and join together. When they are too heavy to stay in the air, they fall to the ground. If it is warm, they will fall as rain. However, if there are violent up and down **drafts,** like in a big thunderstorm, the precipitation may eventually fall as frozen droplets. Those frozen droplets are called hail.

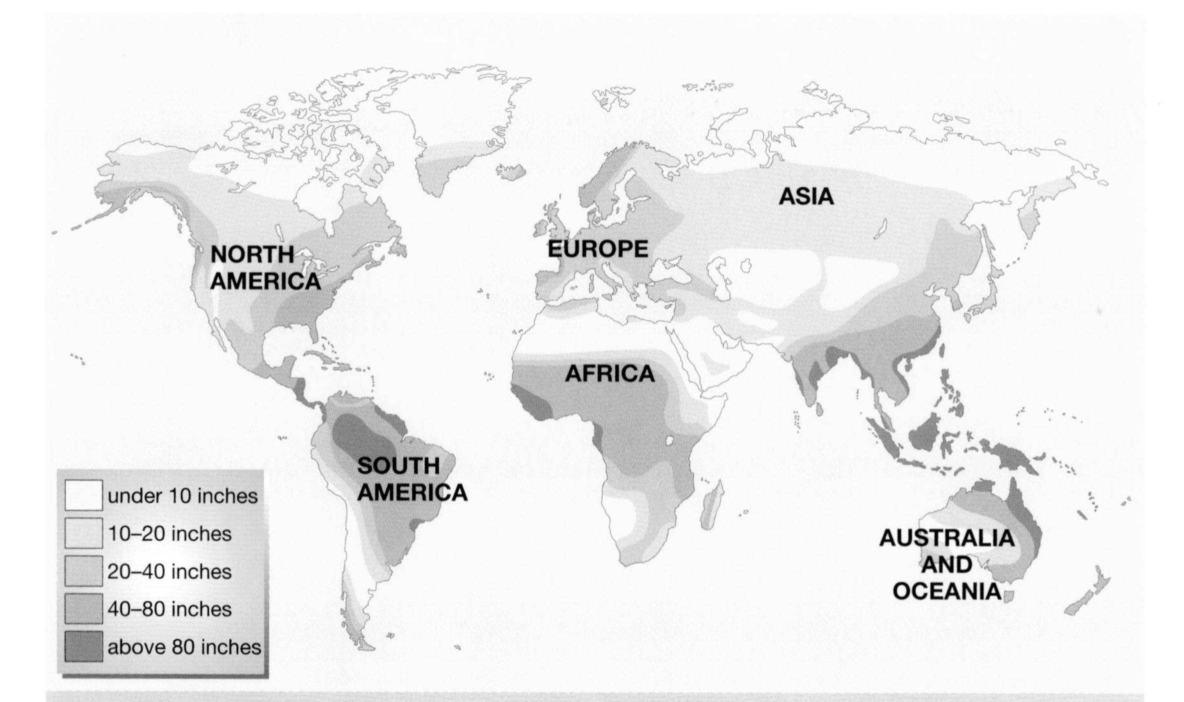

By measuring the average rainfall over a period of time, we see that there is a pattern to where rain falls. This map shows global yearly rainfall patterns.

Certain conditions are needed for precipitation to happen. There must be plenty of moisture in the atmosphere along with rising air. Together, these form clouds. Precipitation occurs when clouds move to a height where the temperature is cold (air cools as it rises because it expands). Clouds can rise in two ways. Sometimes clouds rise up the side of a large object, such as a mountain. In this case, the rain they produce is called **orographic** rain. Other times, warm air currents drive the clouds up to colder parts of the sky. This produces **convective** rain.

Lake effect

Sometimes in the winter, a cold, dry wind blows across a large area of fairly warm water. On its journey, it collects large amounts of moisture. Fog forms above the water. The warmth from the water forces air up, and clouds appear. As these clouds approach the shore and hills, they are forced up even further. This cools the clouds, and so they deposit snow onto places near the areas of water. This phenomenon is known as lake-effect snow, and it often happens near the Great Lakes.

Any large body of water will affect the local **climate.** In the case of lake-effect snow, the snowfall can be very spectacular.

Measuring Rainfall

One of the main instruments used to measure rainfall is a rain **gauge.** The measuring jug inside the gauge has a **scale** that shows how much rain has fallen. The scale must be marked in inches or millimeters. This shows the depth of rain that has fallen over the area.

Making a rain gauge

You can collect useful weather **data** with a simple, homemade rain gauge. You can make one from an empty plastic soda bottle. Try to use one that has a diameter of five inches (127 millimeters), because this is the correct size for a rain gauge. Cut the top off. Although the top is not needed for the rain gauge, you can turn it into an excellent funnel for collecting the rain. Turn the top upside down and place it inside the rest of the bottle.

For more accurate results, an inexpensive rain gauge can be purchased. You can then compare your rain gauge's readings to those given by the purchased gauge.

Professional **meteorologists** use a rain gauge like this one. The outer container stops the water from evaporating. An inner measuring container measures the rainfall accurately.

Be accurate!

- Put your rain gauge where nothing blocks the rainfall.
- Place the rain gauge on a flat surface so that you can read it easily.
- Read the gauge with your eyes level with the scale.
- Measure the depth of the rain collected in inches or millimeters.

It is important that a rain gauge be placed where it collects the most rainfall. Large objects such as trees can block the rain. You can try to find the best location by placing several rain gauges around an area. The results can then be compared. The rain gauge that collects the most rain is in the best place. As a general rule, calculate the height of the nearest tall object. The rain gauge should be located at a distance of two-and-a-half times the height of that object. For example, if a tree is 16 feet (5 meters) tall, the rain gauge needs to be 40 feet (12.5 meters) away from the tree.

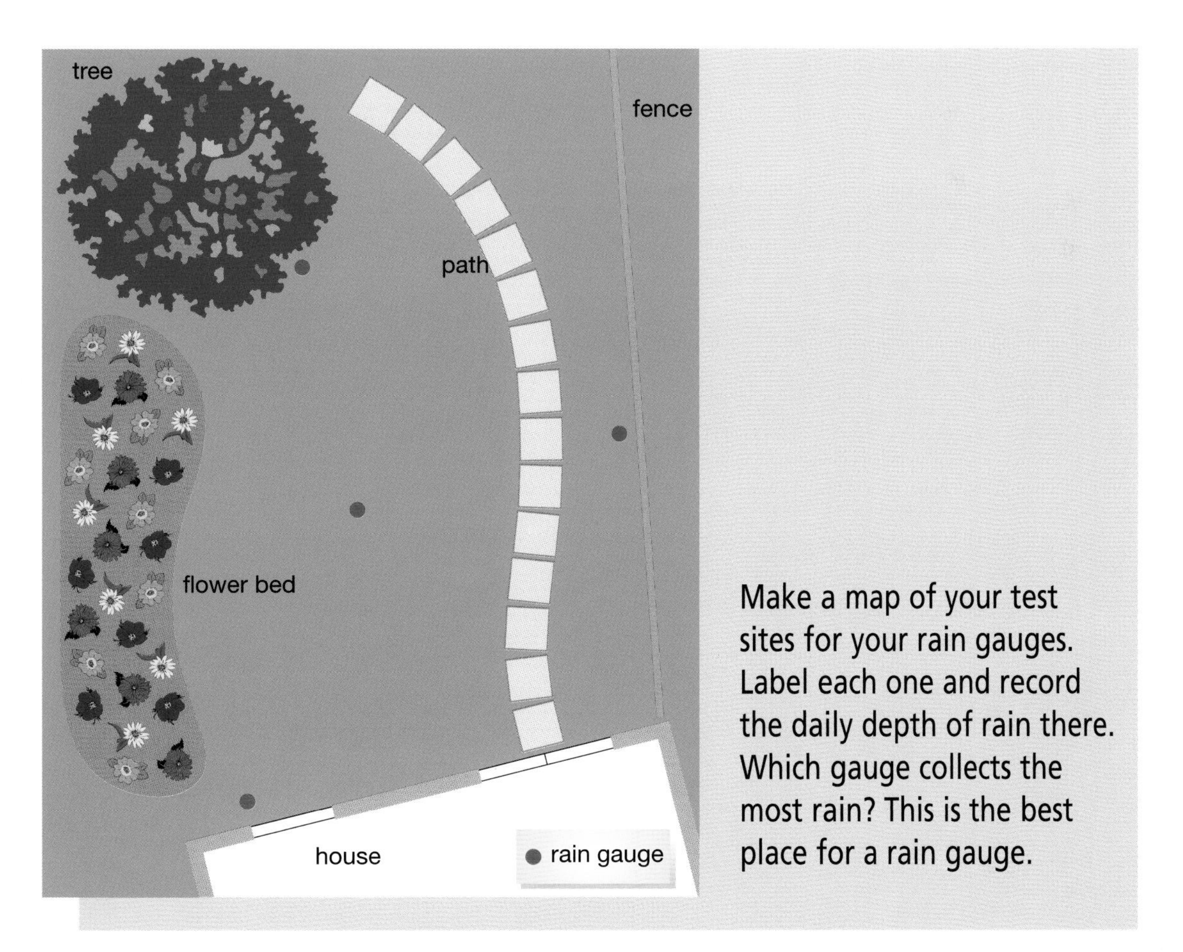

Make a map of your test sites for your rain gauges. Label each one and record the daily depth of rain there. Which gauge collects the most rain? This is the best place for a rain gauge.

Rain and Drizzle

How wet can you get? You may think that **drizzle** is not very wet, but if you spend any amount of time in it you will be soaked, just as if you were in a **downpour** of rain. Both rain and drizzle are forms of **precipitation.** The size and shape of the precipitation, the temperature inside the cloud, and how far rain travels inside the clouds before it falls from the sky all affect the way in which rain falls.

Showers usually come from rounded, **heaped** (**cumulus**) clouds. Drizzle usually comes from low clouds. Rain that falls for a long period of time comes from layered (**stratus**) clouds. The main difference between rain and drizzle is the size of the **droplets** that fall on the ground. The average diameter of a raindrop is around 2 millimeters, whereas the much smaller individual drizzle droplet is 0.2 millimeters. Rain often starts as snow while it is inside the cloud, but as it falls to a lower, warmer part of the sky, it melts and lands as rain.

This downpour is very heavy, but it will only be a shower. If it were drizzling, it would go on for longer. It is worth taking shelter from a sudden shower, as it may soon stop.

Rainbows

A rainbow is one of nature's most wonderful shows. It appears when two things happen at the same time: the Sun shines at an appropriate angle, and droplets of rain fall. As the raindrops fall, they split up the sunlight into the colors of the rainbow (called the spectrum). Lots of raindrops falling at the same time give the impression of a continuous bow of color—the rainbow.

A rainbow can be made outside with a fine spray from a water hose or sprinkler. When the angles of the spray and the Sun are correctly lined up, an artificial rainbow can be seen.

Try this yourself!

You can try this if the rainfall is light.

- Lay out a sheet of light-colored paper in the rain.
- Observe the pattern the rain makes on the paper.
- To make the record permanent, draw around the drops with a pencil before they dry.
- Is there a pattern?

Humidity

Humidity is the term used to describe the amount of water in the **atmosphere. Meteorologists** need to know the water content of the air so that they know whether clouds or fog are likely to form. There will not be rain unless there is a significant amount of water in the air. This varies in different places and in different conditions. The amount of **water vapor** in the air is measured as humidity on a scale from zero to one hundred and is given as a percentage. This percentage is known as relative humidity.

In certain places—like banks, museums, and libraries—low humidity is a good thing, because it is important that paper and books are kept dry. Relative humidity is monitored in case it gets too damp.

There are certain levels of humidity that are more comfortable for humans. We can put up with high temperatures so long as it is dry. If it is hot and the air has a lot of moisture in it (high humidity), however, then it is uncomfortable for most people and animals.

In the deserts of Arizona, there is very little moisture in the air or ground. The relative humidity is very low. Many airplanes are stored there by the U.S. Air Force because there is no need to cover them with waterproof wrappings.

Try this yourself!

Make a simple **hygrometer**—an instrument for measuring relative humidity.

- Use two accurate thermometers to read the same temperature in the same place.
- Mount them close to each other, allowing the air to flow freely around them.
- Tie a square of material to the bulb of one of the thermometers with a piece of thin string.
- Put the loose ends of the string into a small container of **distilled** water.
- You now have a wet bulb thermometer and a dry bulb thermometer.
- Make a note of the readings on both thermometers.
- Look up the humidity on a Relative Humidity Chart like the one below.
- Let's say that the dry bulb temperature is 62°F and the wet bulb temperature is 57°F. Use your finger to trace down the column under 62, and another finger to trace across the row from 57. The relative humidity can be read where your fingers meet. In this case, the relative humidity is 74 percent.

Relative Humidity Chart

Wet bulb Temperature \ Dry bulb Temperature	61° F	62° F	63° F	64° F	65° F	66° F
55° F	68%	64%	60%	56%	52%	48%
56° F	73%	69%	64%	60%	56%	53%
57° F	78%	74%	69%	65%	61%	57%
58° F	84%	79%	74%	70%	66%	61%

Snow and Sleet

Snow is a form of **precipitation.** There are many kinds of snow—from small and dry to wet and large flakes. The size and shape of the precipitation, the temperature inside the cloud, and how far snow travels inside the clouds before it falls all affect the type of snow that we see.

Snow forms when there are ice crystals in thick clouds. As they fall through the clouds, these tiny crystals join together to become bigger flakes. If it is cold on the way down, the precipitation stays in the form of snow. If it is not quite cold enough, there will be a mixture of snow and rain by the time it reaches the ground. Sleet is formed differently. Sleet is made from ice pellets that are formed when rain freezes, or when mostly melted snow freezes.

Snow crystals seen under great magnification have beautiful six-sided patterns. These can include stars, **prisms,** and columns.

Different types of snow

Winter sports enthusiasts appreciate the subtle differences in types of snow. Some types of snow provide better surfaces for skiing on than others. Dry powdery snow falls when it is so cold that ice crystals do not thaw, freeze and join. This is ideal for skiing. However, when fine dry powdery snow falls on top of old snow that is icy and frozen, it does not stick. This means it can slip in huge movements of snow called **avalanches**. These can suddenly sweep down mountain slopes and bury people.

Measuring snow

- Depth of snow is measured with a ruler.
- Use a metal ruler.
- Measure snow that is still smooth.
- Avoid snow that has drifted or been **scored** by the wind and blown into ridges.
- Hold the ruler vertically in the snow.
- Note the reading in inches or centimeters.
- Take two more readings, then take an average (add the three readings up and divide the number by three).
- Clear a fresh area for future readings. If there is more snow you want to be able to measure it from ground level.

It is important to read the ruler when it is vertical. If possible, take the reading near to where you have placed your rain **gauge.** This means that you will be in a clear, open space.

Hail

Hailstones are not made of stone, but are frozen, layered lumps of ice! Most of them are about the size of a garden pea, between five and ten millimeters. In 1970, one particularly bad hailstorm in Kansas produced hailstones that were 7.5 inches (190 millimeters) across. In 1986, a hailstorm in Bangladesh, India produced hail weighing over two pounds (one kg).

Hail can cause damage to property and animals, as well as people. In the 1950s, a lady in Canada wrote to her niece in England:

"Can you imagine a field of wheat two feet [610 mm] high at twenty past six, and at half past six the field looked like a ploughed field."

The whole crop was destroyed. Although hailstorms rarely last more than a few minutes, they can cause an enormous amount of damage.

Even when hailstones are not very big, they can cause terrible damage. This is because they come down with force from a great height. The crop damage seen here was caused in a few minutes.

Hail clouds

Clouds that produce hail mainly form in the strong **updrafts** of spring and summer. In places with very warm temperatures, the hailstones melt before they reach the ground. Hail is formed within giant thunderclouds. Because the air currents are so violent and temperatures vary in different parts of the cloud, the water **droplets** inside the cloud become very cold. As the hail grows in size, it falls, but before it reaches the bottom of the cloud, gusts of wind take it up again. There it gets another coating of ice. As it goes up and down inside the cloud, the hailstone melts and freezes. This continues until it is so heavy that it falls to the ground.

Try this yourself!

- Obtain a hailstone. Keep it cool or it will melt.
- Get an adult to cut it in half.
- Look closely at it using a microscope or a magnifying glass.
- Look at the alternate layers of clear and frosted ice.

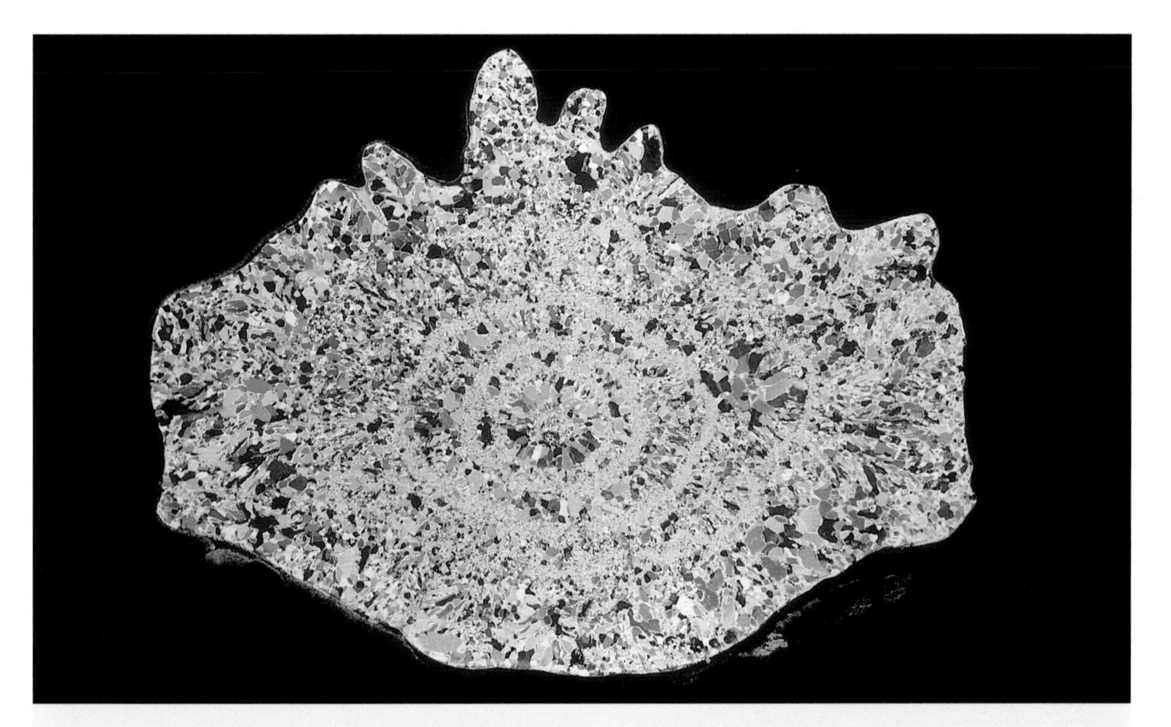

The more powerful the microscope, the more you will be able to see of the alternate layers inside a hailstone. The movement of the hailstone rising and falling inside the cloud causes these layers.

Rain Clouds

There are various sorts of clouds that carry rain. The shape, size, and height of these different clouds can give clues to the kind of weather they may bring, and also help to name them.

Layered clouds let their **precipitation** fall gently, as **drizzle.** This drizzle usually continues for quite long periods of time. Medium-height, layered clouds (altostratus) produce rain only when they are quite thick. Low and dark layered clouds (nimbostratus) bring **downpours** of heavier rain that can last quite a long time.

Dark grey, lumpy clouds release their precipitation energetically, very often as a short downpour. They can blow over quickly, taking the rain with them. The **cumulus** cloud is the main rain cloud of this type. Huge, round, black clouds are called cumulonimbus clouds. These can bring storms with heavy rain or snow. Often there is a noticeable drop in temperature just before it rains.

These towering cumulonimbus clouds stretch up through the sky. They can bring thunder and lightning.

Satellites

Meteorologists use **satellites** to show them exactly where cloud formations are. It is a great help to know well in advance where the clouds are, especially when they are out at sea, where there are fewer **weather stations** to collect weather **data**. The satellite images give a lot of information about the type and behavior of the clouds. By using special cameras, the satellites can see more than an ordinary camera. For example, satellites can see the difference in heat in different parts of the clouds. This may tell meteorologists what is happening in the clouds.

Some people set up their own satellite systems. This can be expensive, but the images are spectacular and give a lot of information about the future of weather patterns. Many universities and weather agencies regularly display their satellite images on the Internet, along with explanations of what they mean.

This color-coded satellite image shows the location of a severe storm. The colors show whether or not the cloud is likely to be dropping precipitation. These images can be used to follow the movement of weather systems in order to make a weather report.

There are many sites on the Internet where you can see satellite images of the weather. Try looking at the web site of a local television news channel, or search for the web site of the national weather service.

Frontal Systems

Frontal systems are shown on weather maps so that people know what weather to expect. When you look at a weather map, you will see special lines called isobars. These lines link together areas with the same **air pressure.** Special lines with curved symbols on them show a warm **front.** Lines with triangular symbols along them show a cold front. A front is the edge of a large body of air. These fronts bring different kinds of weather.

Warm fronts

A warm front approaches with its front edge high in the sky. The following part of the front slopes downward. The first signs of the approaching warm front are the feathery **cirrus** clouds high in the sky that then spread out into **cirrostratus** clouds. Altostratus and nimbostratus follow at the point where the warm front slopes to the ground. These clouds bring rain with them.

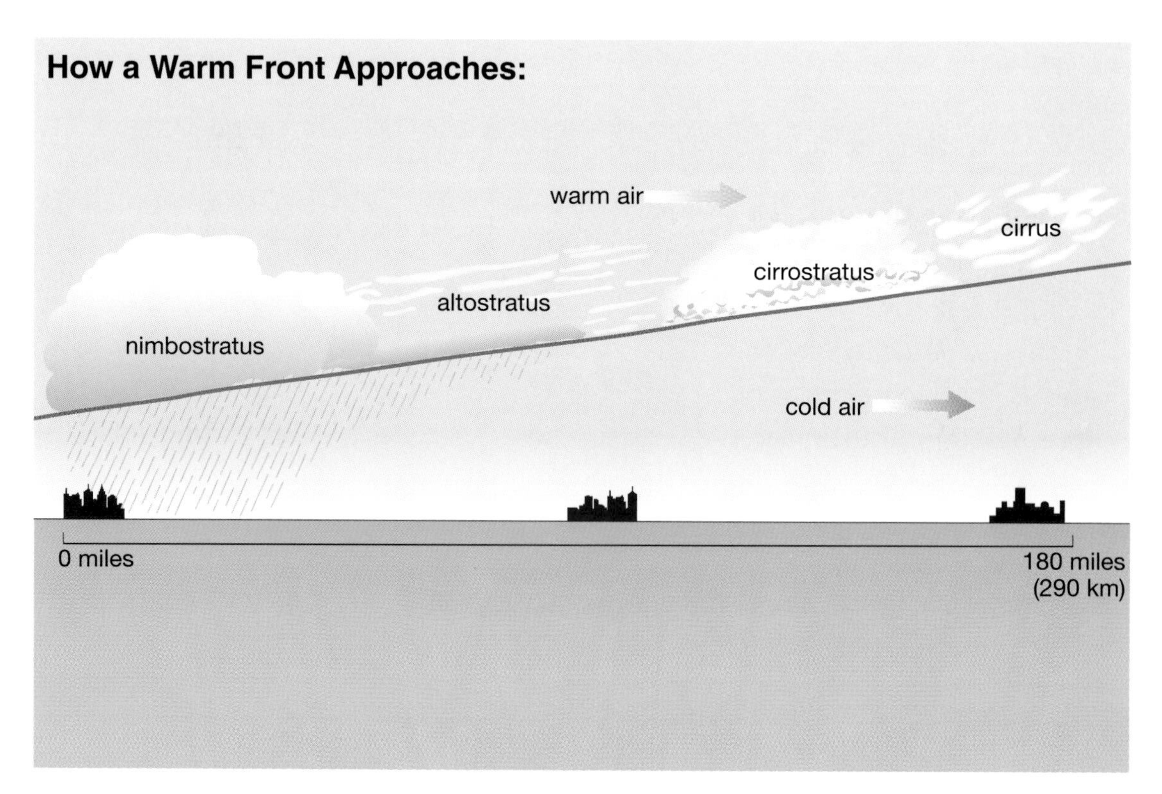

The clouds that this warm front brings can be used to tell when the front will arrive. The dry, cold air pushes the moist, warm air up until it is cooled. When the lower part of the front passes, it will rain.

Cold fronts

A cold front approaches with much less warning than a warm front. It curves up from the surface of the ground, as you can see in the illustration. At first there are **cirrus, cirrostratus,** and **altocumulus** clouds. These are followed quickly by cumulonimbus clouds. Sometimes, cumulonimbus clouds bring **downpours,** thunderstorms, or hail.

You can use your knowledge of fronts to help forecast the weather. By watching for the sequences of cloud types, you will be able to tell when a warm or cold front is approaching. You may see on the weather forecasts on television that a front is approaching. You can then look at the clouds and see how long it will be until the rain arrives. If you see cirrus clouds in a blue sky, you will know that this could be the start of a front bringing rain.

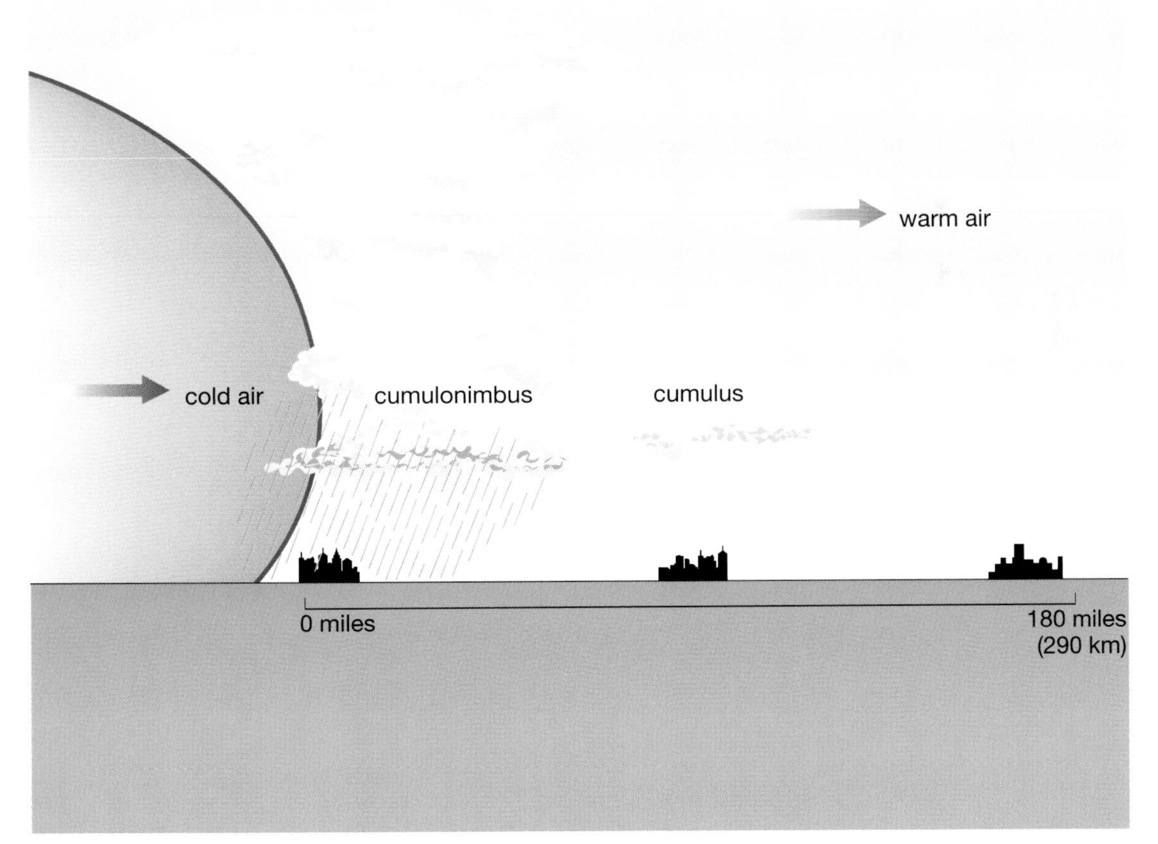

The front edge of the cold front is a very different shape than the front edge of a warm front. The cold air, pushing into the warmer air, causes cumulonimbus clouds to appear with a lot of **precipitation.**

Microbursts

Microbursts are dramatic weather events that cause strong bursts of wind. Microbursts often occur as a result of isolated rain showers or thunderstorms. There are two sorts of microburst: wet and dry. Wet microbursts occur in wet conditions and dry microbursts occur in dry conditions.

In a dry microburst, a column of rain suddenly falls from a cloud into the dry air beneath. The rain quickly **evaporates.** This evaporation cools the air. The cool air is heavy and sinks, making a powerful downward wind under the cloud. The wind hits the ground and spreads out, traveling about another 2.5 miles (4 kilometers). This wind might last only from five to twenty minutes, but it can reach speeds of up to 134 mph (215 kph). However, a distant observer may only notice a dry microburst disturbing the dust on the ground.

Wet microbursts happen when it is raining very heavily. In these microbursts, a lot of the rain reaches the ground along with a **downdraft.** The wind and rain hit the ground with such force that they produce another wind that goes out sideways. This sideways wind is called windshear.

In this wet microburst, you can easily see the **downpour** of rain that causes the strong burst of downward wind.

A dangerous kind of weather

Microbursts can be dangerous for airplanes, especially when they are landing or taking off. The powerful winds they produce can cause airplanes to have serious problems.

Work is being done on predicting microbursts. If the temperature at ground level is very different from the temperature in the air, this means that microbursts are more likely. Two indexes are being developed. These are systems that will help to predict when the weather conditions needed for the two types of microburst are present. They are the Dry Microburst Index (DMI) and the Microburst Day Potential Index (MDPI). They are currently only being used in certain parts of the United States of America.

Knowing all about the local weather at airports is very important. As they prepare for takeoff or landing, information about the weather is passed on to the pilots, and they can then make decisions about the effects of the weather on the safety of the plane.

Rain and Snow Warnings

Rain and snow have a big effect on people's lives, so it is important to know when they will happen. Professional **meteorologists** try to gather a lot of information about the weather from as wide an area as possible. They can warn people about the possibility of bad weather. This is very important when there is going to be a lot of **precipitation.** When monitoring the weather, meteorologists share the **data** they collect, and pass on warnings to each other.

Some weather watchers use devices called **radar.** The radar bounces special waves off the clouds into a receiver. These show where a storm is and which way it is traveling. **Satellites** in space can also spot a storm developing out at sea. As a storm comes closer inland, radar stations can continue to follow it. It is important to try to predict serious weather conditions. On August 14, 1975, a storm in London, U.K., surprised people with almost seven inches (170.8 mm) of rain in two and a half hours. One person drowned and cars floated down the roads. If this violent thunderstorm had been predicted, the huge downfall of rain may have caused less damage.

Regional, national, and international weather forecasts begin with readings taken at **weather stations** in many different locations. This weather station has its own radar equipment, which is housed inside a special dome.

Hurricanes, floods, and snow

Hurricanes are tropical storms. They can bring huge amounts of rain that can cause terrible damage. Meteorologists receive hourly satellite pictures. If thunderstorms look as though they are developing into a hurricane, satellite images are used to look for signs of rotation. Hurricanes are monitored and warnings are passed to ships, aircraft, and the general public.

Flood warnings can be very helpful. One way of predicting flooding is to study rainfall at the starting points of rivers. Measuring stations can be placed along rivers. They note the amount of water flowing by and measure the water in all parts of the river. This **data** helps to predict flooding further down the river.

Snow also needs to be monitored. If a lot of snow falls in the winter, it could cause flooding when it melts. Snowfall is also monitored to see if it is stable. This means it is not likely to cause **avalanches.** These can destroy whole villages.

If people are warned early enough about bad weather, they can make preparations to reduce damage to life and property. Here, residents of North Carolina prepare for a hurricane.

Floods and Drought

The term "drought" means that no rain has fallen where it would normally be expected to fall. The amount of rain expected depends on the **climate** of the area. A dry summer for one country would be considered a wet one for another country. For example, rain normally falls all year round in Europe. If fifteen days go by without rain, or if it rains less than .02 centimeters (0.2 mm) per day, a drought is declared.

Very often in the summer months, an area of high **air pressure** will stay over a region and will not move. While this high air pressure stays still, the normal **depressions** that bring rain are forced into different areas. This situation happened in 1976 in Europe. The northwest of Europe had hot, dry weather while the Mediterranean had a long, wet summer!

Bringing relief

For many people in Africa, long periods of drought bring suffering. Crops cannot grow and water supplies dry up. In places that suffer long droughts, attempts can be made to encourage the formation of rain clouds.

Research is continuing into the best ways to bring rain to areas that need it. These airplanes are trying to cause **precipitation**—you can see the chemicals one of them is spraying.

One of the ways in which clouds naturally form is from water **droplets** sticking to tiny particles in the **atmosphere**. If there are not enough of these particles, airplanes can spray particles of harmless chemicals into the clouds. The droplets then form around these particles, and rain may follow. However, **water vapor** must already be in the air for this process to work, so it is not very reliable.

There are lots of countries in the world that switch between having hot, dry weather and warm, wet weather. These countries rely on receiving a lot of their annual rainfall in a short period of time. The people of India call this great rainfall period the monsoon season. This is a seasonal wind system that brings wind and rain. This arrives in May and continues for six months. Its arrival breaks a long, hot, dry spell and brings relief to millions of people.

This heavy rain is called a monsoon. People are usually pleased when it comes, because it ends a long spell of dry weather.

Glossary

air pressure pressure at the surface of the Earth that is caused by the weight of the air in the atmosphere

altocumulus type of cloud found in patches in the middle height of the sky that are usually white or grey. This type of cloud is shaped in rounded heaps.

atmosphere gases that surround our planet. They are kept in place by the planet's gravity.

avalanche huge amounts of snow, with ice and rock, which travel rapidly down the side of a mountain

cirrostratus type of cloud found very high in the sky. It is thin and transparent and made of ice crystals.

cirrus highest form of clouds, made up of ice crystals in thin, feather-like shapes

climate weather conditions in a place over a long period of time

condense when water vapor returns to its liquid state

convective vertical movement, especially upward, of warm air

cumulus kind of cloud, consisting of rounded heaps with a darker horizontal base

data group of facts that can be investigated to get information

depression area with low air pressure readings (a cyclone, for example)

distilled pure water that has evaporated and condensed again

downdraft current of air moving in a downward direction

downpour heavy fall of rain

draft small current of air, often caused by air being let in through an opening

drizzle small, light rain

droplet little drop

evaporate when water changes from a liquid into water vapor

freezing rain rain that passes through a cold layer of air and then freezes when it hits the ground or other objects

front front edge of an air mass, where it meets air of a different temperature

frontal system area where air masses of different temperatures and humidity meet

gauge measuring equipment, in this case used to measure rain

heaped piled up in mounds

hydroelectric electricity produced by means of water power

hygrometer instrument that measures the relative humidity of the air

matter material that things are made of

meteorologist somebody who collects weather data and studies the weather

orographic related to mountains

polar region area surrounding either the North Pole or the South Pole. Both poles have very cold climates.

precipitation moisture or water vapor that condenses and falls as rain, hail, or snow

prism object or material that splits light into separate colors

radar use of radio signals to find out about objects, including how far away they are

satellite manmade device that orbits around the Earth, receiving and transmitting information

scale numbers used to represent measurement on instruments such as thermometers

score to mark with a notch or incised groove; in this case, one made in the snow by the wind, which makes the snow uneven

stratus forming a layer

updraft rising current of air

water cycle the cycle in which water from the sea evaporates into the atmosphere, condenses, and falls to Earth as rain or snow. It then evaporates directly back into the atmosphere or returns to the sea by rivers.

water vapor water in the form of gas

weather station collection of weather instruments that measure the weather regularly

More Books to Read

Bundey, Nikki. *Snow and People.* Minneapolis, Minn.: Lerner Publishing, 2000.

Chambers, Catherine. *Floods.* Chicago: Heinemann Library, 2000

Gardner, Robert. *Science Project Ideas about Rain.* Berkeley Heights, N.J.: Enslow Publishers, 1997.

Index